Leadership Development - Professional Skills to Catalyze Your Career

AA009101

Topical Conference at the 2022 AIChE Spring Meeting and
18th Global Congress on Process Safety

San Antonio, Texas, USA
10 – 14 April 2022

ISBN: 978-1-7138-6823-1

Printed from e-media with permission by:

Curran Associates, Inc.
57 Morehouse Lane
Red Hook, NY 12571

Some format issues inherent in the e-media version may also appear in this print version.

Printed with permission by Curran Associates, Inc. (2023)

For permission requests, please contact AIChE
at the address below.

AIChE
120 Wall Street, FL 23
New York, NY 10005-4020

Phone: (800) 242-4363
Fax: (203) 775-5177

www.aiche.org

Additional copies of this publication are available from:

Curran Associates, Inc.
57 Morehouse Lane
Red Hook, NY 12571 USA
Phone: 845-758-0400
Fax: 845-758-2633
Email: curran@proceedings.com
Web: www.proceedings.com

Leadership Development - Professional Skills to Catalyze Your Career

Topical Conference at the 2022 AIChE Spring Meeting and 18th Global Congress on Process Safety

San Antonio, Texas, USA
10 – 14 April 2022

AIChE
The Global Home of Chemical Engineers

TABLE OF CONTENTS

641119 Climate & Policy Context

Abstract:

Come provide your input to policy priorities for AIChE on climate solutions across energy, transportation, renewables and other areas. This session will build on The Climate Solutions Community workshop in November in Boston and feature roundtable breakout discussion to identify where policy engagement with government can help accelerate solutions. Broad input from members, Divisions and Forums is desired.

639348 Finding the Leader within You: It's a Journey, NOT a Destination!

Abstract:

"Leadership" is a term that is often overused and, at least as often, ill-defined and misunderstood. Nevertheless, the current climate within the Process Industries (PI) is crying-out for real leadership, yet it seems to be in egregiously short supply (as well as substituted by a deplorable excess of "management").

This presentation will review many of the lessons learned, the "course corrections" and resources used by a 30+ year veteran of the PI that have helped him discover and develop his potential. Unlike many step-by-step or "paint-by-numbers" formulas that one finds promulgated in the personal development marketplace, the author will instead share his experiences and perspectives, gleaned over the course of his career, which one may use as a framework for discovering and unlocking one's own leadership potential. In addition, the author will present his perspectives on how his personal journey has, and continues to, affect his career trajectory in unanticipated ways.

639292 Admitting Human Error in the Workplace - a Case Study in Maximizing Safe and Reliable Performance with Human Factors

Abstract:

Human error is inevitable and serves as a symptom of systemic failure. As with all symptoms, it becomes a signal spurring systemic change. The idea is to welcome error as an indication that leadership should investigate the dark, murky, uncomfortable cultural issues that made the error seem like a good idea at the time. Instead of blaming individuals for error, ask why the system responded the way it did. In this interactive session, participants will use an on-screen response system to experience a developing investigation as facilitators of a real-world example. We will study an organization's emergent behavior through the actions of two operators. As a rule, these operators did not set out to cause failure; rather, they intended to contribute to the larger success of the organization. Nevertheless, as part of a multi-dimensional operating framework, they were shepherded by complex events into making decisions that instead caused an incident.

OPERATIONAL EXCELLENCE AGILITY™

APPLIED TO THE OIL & GAS INDUSTRY

Curtis Bakle, Palmetto Premier Consulting, Charleston, SC

Jeff Gill, Palmetto Premier Consulting, Houston, TX

Introduction

We are in the middle of a major disruption in the oil, gas, and chemicals industry where the demand for fossil fuels is decreasing and the refiners and processors with the finest pedigree are winning. Those companies that had the advantage of starting a digital transformation before the COVID-19 pandemic now have the leading edge, but this disruption has caused many programs to stall.

In late 2020, McKinsey & Company completed a survey of 2,190 global respondents across many industries. Only 44% of those respondents reported an agile transformation in their company and two-thirds of those companies said, "that their organizations were just treading water."2 Agile is not limited to the IT industry or departments. Rather, it is a tool that delivers value across many teams and all industries. For the oil, gas, and chemicals industry, only 47% of the industry claimed to have an agile transformation. The agile success stories are plentiful and average a 30% improvement on efficiency, customer satisfaction, employee engagement, and operational performance. In the past, it was easier to focus on building empowered workforces and resiliency of Operating Management Systems (OMS), which led companies to use resources to identify risks and manage programs that delivered safe, reliable, and compliant operations.

The "old normal" has not been replaced, but added to the mix is a Volatile, Uncertain, Complex, and Ambiguous (VUCA) environment that has many companies juggling priorities with a disrupted workforce that is burnt out and not at their peak collaboration. Companies should design for disruption and respond quicker if they hope to crush the competition and win on Wall Street. Operational Excellence Agility™ (OpXL Agility™) is creating a focused, fast, and flexible mindset that delivers systematic operations at commitment levels that continuously challenge the process to deliver change. There is a strong passion for improvement and setting new targets in response to VUCA environments.

Operational Excellence Agility™ Defined

Operational Excellence Agility™ is achieved when the three major components of an organization are in balance: operational excellence, people and culture, and agile leadership. When the performance is not in balance, companies will not reach their full potential and beat their competition. Most companies typically perform very well at one or two of the components, not all three. The application of agility to the three major components creates a mindset that embraces a VUCA world and is prepared for the next disruption.

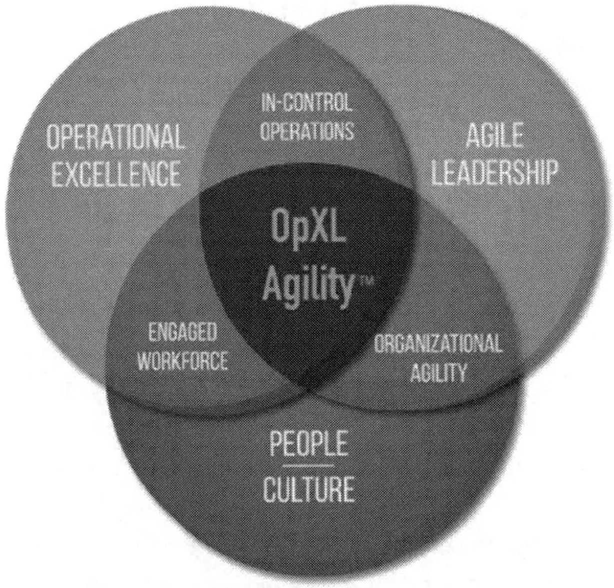

1. **Operational Excellence** is combining operations, knowledge, and experience to drive change in the process to achieve safe, reliable, and efficient outputs.

2. **Agile Leadership** is the ability to glean insights from the data, engage the team, and take action to improve the outcomes.

3. **People and Culture** is building a collaborative team that knows their strengths and weaknesses with a bias to deliver even higher results.

Ensuring the programs are systematic, including a verification of the activity, creates a vision of these gaps and can apply a priority for action. At the same time, leadership tends to press on corrective actions and hyperfocusing on the details of the actions. As a result, they lose focus on their role in leadership, as well as the need to anticipate change. In our experience working with large companies, many feel that their processes are well developed and routine. However, to survive in competitive industries, companies are cutting costs. The Operational Excellence Agility™ model presents the balance of the process, people, and leadership needed to make the program consistent enough to operate businesses. These principles reflect the selection of the right team with a continuous improvement mindset that challenges them to identify gaps, as well as listen to customers that have changes coming in the pipeline. This allows them to work preemptively to anticipate the change before causing a disruption in the system. In the current VUCA environment, it is difficult to predict or fully understand the extent of the change, however many probable disruptions can be predicted. Supply chain disruptions, climate change, environment, social and governance (ESG), technologies, and the COVID-19 pandemic are a few examples that have occurred in the last three years. This model creates a mindset to be focused, fast, and flexible, which allows businesses to respond quicker than the competition and take advantage of the potential value.

Operational Excellence

Operational excellence processes drive the industry to constantly improve activities of the OMS program. These programs have been well established for over 15 years at the larger companies, creating a framework or foundation for a systematic and consistent approach to managing operations. With a mature process and resources, this operation can become complacent, showing little improvement and change. A verification process can ensure that the monitoring of the programs is kept at the forefront of the production meeting and be a source of data for setting future targets and improvements.

The Operational Excellence Agility™ program also provides assessment. While working with a small startup, where operational excellence is still being developed and processes are not yet systematic, gaps in availability, reliability and maintenance can lead to cost inefficiencies. The Operational Excellence Agility™ program was applied along with the leadership team to create a cross-functioning priority of actions. The result focused on operating procedures, power consumption, competency development, and reactor outage timing. The reactor outage project reduced the time to prepare the shutdown of the reactor system by 33%, which further increased availability.

Many companies operate with several process improvement action items. Sometimes several is too many. These teams can be caught in the "Too Busy Trap" when they feel overwhelmed by the number of actions, the priority of the actions, or the complete effects of their value. When combined with a complex mix of processes, audits, reports and evaluation sessions, the workload becomes increasingly unmanageable. Breaking this cycle in the Operational Excellence program only requires a few adjustments for teams to gain the full value and priority of process improvement actions. For example, while working with a team that was inundated with actions, we simply changed the vision to a list called "Be the Best". Setting the vision, priority, timing, and funding helped align the team and let them take ownership of tactics that were well planned out for many months.

Agile Leadership

Defining agile leadership is not always easy. Most people come into a role without having the fundamentals of leadership. They may have performed spectacularly as an individual contributor in their previous role, but now must manage that same team they were a part of, which is a different skill set entirely. Many people can develop into great leaders, understand what they have to do to create value, manage a team, or deliver a project. They have the foresight to develop a team that can fully utilize their skills to create a collaborative organization. The ability to be an agile leader is another tier of expectation altogether. This is about how you show up as a leader: by creating a strong bond of trust with their team, striving to anticipate change, and creating a mindset that will challenge the team to be focused, flexible, and ready to pivot with the business environment.

These skills translate onwards to the team. A good leader with agile leadership skills can develop a team that is not afraid to tackle the hard-to-do list. This also reduces the barriers associated with having just one strong leader and enables the creation of a strong agile leadership team. This highly collaborative team can glean insights from the data to forge change and improve outcomes. The right data and the right leadership mindset are only part of the Operational Excellence Agility™ model. People and culture have to be willing to set aside biases and strive for continuous improvement.

People and Culture

An engaged workforce with great leadership can take the performance of a team to the next level. You can still have the engaged workforce without a leader so long as no negative influences are applied to the team, but the team may struggle to align with overall business strategy. Having engaged workforce that has a great agile leader creates an organization that knows their strengths and weaknesses and can deliver challenging results.

To achieve this, you need to continuously pump energy into assessing your culture, ensuring it takes time for culture to become an unconscious habit. Leadership has to set the stage and be the example for people to follow. Each individual is accountable for their behavior and understands their role in the team. Many interviews have focused on the technical skills of the person without inquiring as to how they might fit into the team. Diversity is one of the keys to having a high performing team: not just diversity of gender but diversity of thought, too. Many companies will evaluate candidates not just on their technical skills but also their personality, teamwork, empathy, and leadership.

Operational Excellence Agility™ Conclusion

Operational Excellence Agility™ is creating a mindset that applies a level of thinking that is focused, fast, and flexible. This mindset delivers systematic operations with people that challenge the process, leading to improvements in all levels in the organization. The balance of Operational Excellence, people and culture, and agile leadership, will have a team achieve peak collaboration, and allow you to crush your competition by responding to the VUCA environment quicker. The model is not limited to the oil & gas industry and has applications in many businesses in many industries.

References

1. Horney, Nick and O'Shea, Tom (2015). Focused, Fast, & Flexible: Creating Agility Advantage in a VUCA World. Indie Books International.
2. Aghina, Wooter, Handscomb, Christopher, Salo, Olli, and Thaker, Shail (2021, May 26). *The impact of agility: How to shape your organization to compete.* McKinsey & Company. https://www.mckinsey.com/business-functions/our-insights/the-impact-of-agility-how-to-shape-your-organization-to-compete.

639925 Why Leadership Development Fails: A Lesson for Our Industry.

Abstract:

According to one source leadership training is a $366 Billion global business and 95% of learning organizations either plan to increase or maintain their current investment in leadership development. At the same time, many CEOs complain that they do not have enough effective leadership in their organization to bring about the change and level of performance needed.

In this session I will discuss the top reasons why we are not successful with attempts to develop effective leaders in our industry. I have experienced each of these over my 40 years in the chemical industry. Now I am teaching leadership to undergraduate engineering students in attempt to get in front of this barrier to success.

This session will give organizations and individuals a fresh look at how to unlock the hidden leadership potential in their staff and themselves.

640983 Making the Most Out of AIChE as a Young Professional (YP) Member

Abstract:

AIChE offers many benefits and opportunities for early-career chemical engineers, including networking, professional development, leadership, and educational opportunities. In this session, we will provide an overview of opportunities to grow your chemical engineering career as a young professional through engagement with AIChE. This session includes a discussion on AIChE structure, highlighting opportunities to get involved with groups and communities that meet your needs and interests. In addition, we will cover the Young Professionals Committee (YPC), a group of volunteer-YP members that plan programming and resources for early-career engineers year round. Further, we will dive into tips and tricks to make the most of attending AIChE Meetings (Spring and Annual) as a YP member, including highlighting some of the feature YP events like the YP Socials and Annual Beer Brewing Competition (Fall). We will conclude with a panel discussion featuring AIChE YP members past and present discussing how AIChE involvement has positively impacted their careers.

641599 Women in Chemical Engineering Community (WIC) Breakfast and Keynote

Abstract:

Women in Chemical Engineering Community (WIC) Breakfast and Keynote

The Women in Chemical Engineering Community (WIC) invites you to attend this interactive breakfast event. This networking event aims at bringing WIC members together in an informal setting. Members will have the chance to network together, exchange ideas and share knowledge while having a buffet style breakfast on round tables. During breakfast, WIC is honored to have Dr. Cristina Thomas give WIC members a keynote presentation covering impactful leadership from her chemical engineering and management career as well as becoming a STEM Equity trailblazer. Dr. Thomas is an amazing industrial leader at 3M. Dr. Thomas's research accomplishments are inspiring. Her abilities to fuse innovations with business strategies and commercialization are intriguing. In addition, Dr. Thomas has always been an AIChE volunteer, particularly with WIC, the Management Division, PAIC and the K-12 Community. We welcome you to this event as we all learn from Dr. Thomas inspiring lessons on integrating life and career while leading. After the keynote, attendees will have a chance to ask Dr. Thomas questions.

Meet our keynote speaker

Cristina Thomas

Dr. Cristina Thomas has more than 30 years of experience in R&D organizations including in the Venezuelan oil industry, IBM (New York) and 3M (Minnesota). She has degrees in Chemistry and Mathematics and a Ph.D. in Chemical Engineering. Her 3M assignments have included Global R&D Services Senior Director, R&D Global Process Owner, Strategy Manager, Laboratory Manager, Commercialization Manager, Design for Six Sigma (DFSS) Technical Black Belt and Group Leader of the Materials Modeling Group.

She became an early leader in the molecular and materials modeling community (AIChE's CoMSEF Forum). At 3M, she led the invention of 3M window-film products, of pavement markings retroreflective solutions and of connected safety platforms. She is a coinventor in 20+ patents. She received the 2019 AIChE Industry Leadership Award, the 2019 AICHE Management Division Award, and is an ACS fellow and an ACS PMSE fellow.

She has a passion for moving new technologies into markets and for education of underrepresented students and women in science and engineering and co-leads 3M STEM and Skilled Trades initiatives. She is a member of the AIChE Corporate Council & Industry Leader's and the Government University Industry Research Roundtable (National Academy of Science). She lives in Stillwater, Minn., with her husband Scott, her dog Dulce and three adult children.

641598 Women in Chemical Engineering Community (WIC) Panel

Abstract:

The Women in Chemical Engineering Community (WIC) invites you to attend this interactive panel. Inspiring, successful, engaged, and experienced women leaders from academia and industry will share their journeys with us. Women navigate unique paths throughout their journeys that impact their careers and lives and those of others in many different meaningful ways. Come listen to these journeys and get motivated and inspired. The panel will start with short introductions from our panelists. The floor will then be open to attendees to ask questions to our panelists. The panel will conclude with words of wisdom from our panelists.

Saadet Ulas Acikgoz has been working for Honeywell UOP for the past 15 years. She is currently the Senior Manager of R&D Process Modeling and Optimization group. Prior to that, she held several specialist and management positions within the UOP R&D organization where she was involved in scale-up and commercialization of several UOP process technologies, development of new tools to improve experimental efficiency and commercial data analytics. During her career at UOP, she co-authored more than 15 UOP patents and patent applications.

Saadet first joined the AIChE as a student member and she has been involved in the programming of AIChE Spring National Conferences since 2009, serving in roles as Session Chair, Co-Chair and Topical Chair. She co-authored and presented several papers in the AIChE Spring National Conferences and the Annual Fall Meetings since 2004. She served as a 3 year director, treasurer, second vice chair, first vice-chair, chair and past chair for the Fuels & Petrochemicals Division (F&PD). She also chaired the Annual Meeting F&PD programming and now serving as the vice-chair for the F&PD Spring Meeting programming.

Saadet is also serving on the AIChE Executive Board of the Program Committee and the AIChE Chemical Engineering Progress Editorial Advisory Board.

Saadet received her MS and PhD Degrees from the University of Illinois at Chicago in Chemical Engineering and Bioengineering with a specialization in process systems engineering. She holds a BS Degree in Chemical Engineering from the Middle East Technical University in Ankara, Turkey.

Wendy Young is the Sales & Marketing Manager for Chemstations, providing process simulation software solutions through the CHEMCAD product suite. Her experience includes working with technology developers and users in a broad range of industry areas: chemical and pharmaceutical manufacturing, refining, and water treatment. Wendy obtained her B.S. degree in chemical engineering from Texas A&M University (1997) and her MBA from Rice University (May 2020).

Wendy has demonstrated commitment to AIChE by supporting local sections (founded YP Group of South Texas Section), growing membership (Chaired Membership Committee), and donating to the Foundation (Silver level). She's privileged to work with colleagues across industry, academia, and government through her past service as Chair of the Young Professionals Advisory Board (now YPC), on the Board of Directors (2007-2009), and Chair of CTOC (2017). Wendy advanced AIChE's leadership in addressing global challenges as a founding Co-Chair of AIChE's Water Initiative (now ISWS). She was the Co-MPC for the 2015 AIChE Spring Meeting and was elected to Fellow in 2016.

She was elected as the AIChE Treasurer for 2021 through 2023.

Jennifer Mize is passionate about protecting people, property, and the environment through the application of risk-based process safety. Over the course of her career, she has developed expertise in relief design and risk assessment and is committed to sharing her expertise with others as a supervisor and mentor and through her volunteer commitments with the Center for Chemical Process Safety (CCPS). Jennifer is currently the manager for the Process Safety Design group at Eastman. She is in her first year of a 3-year director term for the East Tennessee Local Section (ETLS) and has enjoyed serving in several other roles for the local section during her time at Eastman, including serving as Chair in 2003. At the Institute Level, she is serving as Second Vice Chair of the Safety & Health Division. Within CCPS, Jennifer is the Chair of the Process Safety Incident Database (PSID) Steering Committee and a member of the CCPS Process Safety Professional Certification (CCPSC) Managing Committee and Exam Committee. She is the 2022 Vice Chair of the Process Safety Management Mentoring (PSMM) Conference in the 18th Global Congress on Process Safety. In addition to her support of AIChE, she is a contributing member of the ASME Section XIII Committee on Overpressure Protection.

Jennifer graduated from Auburn University in 1996 with a bachelor's degree in Chemical Engineering. She is a registered professional engineer in Tennessee and holds CCPSC certification.

640772 Defining Your Superpowers-strength based growth

Abstract:

Like every superhero, you have defining core strengths that when realized and enhanced help you be successful in complex roles and opportunities. Your journey starts by becoming self-aware and realizing your true potential. The question remains on what your core strengths are and how do you identify those. This session aims to share 1) tools that you can use to identify your strengths, 2) techniques that can help you enhance your strengths and identify the blind spots that can be detrimental and 3) case studies defining alignment of strength and job needs. We will use an example tool to help audience understand how these assessments are done with follow-up questions intended to drive more engagement from the audience.

640775 Art and Science of Stretch Assignments

Abstract:

Stretch assignments help you develop expertise or gain experience outside of your comfort zone or current role. By their very nature they tend to be challenging, hence it is important to approach these opportunities in a strategic way that will help you succeed. These assignments are tools that can help you grow and, in some instances, accelerate career progression if executed with diligence. In this session we will cover how to advocate for yourself, be put on a short list, determine if a stretch assignment is a good fit, and potential ramifications of poor performance. We also will learn about case studies and personal experiences, including both successes and failures.

640773 Foundational Skills for Career Growth

Abstract:

This talk will describe the foundational skills that enable career success, and share insights on how to acquire and develop these skills. In addition to the core technical competencies required for a job role, it is critically important to develop a suite of skills for networking, problem solving, communication with the multiple modes available and project management. This is just a starter list for the variety of ancillary skills needed to thrive in today's work environment.

640866 IDEAL Leadership Panel

Abstract:

In keeping with AIChE's recent adoption of the IDEAL Statement, this session will bring together Leaders in Industry and Academia that embody the IDEAL Principles. With a focus on the L for learning, these leaders will offer critical perspectives from the chemical engineering community to address the importance of early adoption of equity, diversity and inclusion efforts in an all settings so that these principles can take root and last in the professional setting. The panel will address questions such as different skills leaders need in an IDEAL world, why IDEAL is important to everyone, not just those in marginalized groups, the benefits of Employee Resource Groups and how to support them, and sponsorship of IDEAL communities so they don't just survive but thrive.

638772 Leadership and Culture - the Key to Success

Abstract:

Culture plays an important role in how a company operates since culture is all about a company's values, beliefs, symbols and myths, which have developed over the history of the company. Culture can play a significant role in a company's ability to succeed. A company that fosters an open, honest and non-punitive culture, encouraging, fostering and rewarding initiative and success, is a much better place to work as opposed to one that has a closed, feared and punitive culture; these organizations suffer from a lack of initiative and employees just doing as they are told. Culture stems from the highest levels of the organization and is reflected in how the CEO works with the management team, all the way down the line to local supervisors; it dictates the norms of behavior within the organization. Leadership and management style is therefore very relevant in how the culture of the organization develops. Creating a culture that facilitates personnel growth and development, at the same time as providing a great place to work, to enhance the company's success, is crucial. This paper explores the leadership styles required to enhance success.

640193 Empowerment of Employees Motivation

Abstract:

Leadership in an organization describes the different skills required in a leader and the types of leadership required in different situations, such as risk-taking and risk handling ability, self-confidence, self-concept, creativity, and strategic thinking. All these trails are important in determining how to catalyze your career. "Character" is the first leadership concept that provides moral authority to bring people and resources together. Character gives the persistent spirit to pursue a goal despite a lot of hindrances and obstacles. This is the most essential leadership concept to pursue a vision and mission, and to move an entire enterprise and enhancement approaches to employee motivation. In this topic, I'm going to draw my opinion on how to empower the employees' motivations during work and how to build relationships with customers and offer a product or service to meet their needs. Improving service quality and trying to increase my efficiency to increase the value to the customers is one of the best steps to creating a satisfaction average level for customers. Also, confidence is the next leadership concept that I must develop. This is essential to boosting and motivating me in my daily work.

639052 Lead It! Solve It! Few Tips, Tools, and Strategies for Becoming an Effective Technical Project Leader from a Slightly Seasoned Professional

Abstract:

Most of the young inexperienced engineers and scientist enter the workforce without grasping the reality that at some point in time in their careers they will have to take on project leadership roles. This session will focus on technical projects and how to manage and lead them effectively. Some of the most common tools and strategies to solve complex technically challenging process problems will be presented as well as the strategies on how to deal with people issues. The importance of networking, recognition and project closing steps will be discusses as well.

639100 How to Work Best with Non-Engineers: Real Advice from a Real Operator

Abstract:

Have you ever formulated the best engineering solution to an operational issue? Used all of your engineering knowledge to solve a problem that will revolutionize the way that your facility will operate from now until the end of time? Only to walk into your control room to explain it to the Process Technicians or Operators, and they look at you like you are from another planet? The communication gap between the "thinkers" and the "doers" in an industrial facility is a common trope, but one rooted in truth. These misunderstandings generally result from a simple language barrier, body language, work history issue - or a combination of all three. Taking the time to understand the mindset of a Process Technician or an Operator will pay dividends in the execution of the next big idea you come up with. Merging the worlds of the "thinkers" and the "doers" at your facility can be done if you take a little time to understand your audience and what motivates them.

When a portion of the plant is failing or not running efficiently, it generally is due to faulty equipment, technology issues, or budget constraints. As an engineer, you are charged with making the plant money and doing so safely. Consider this example: you have identified a problem with an ethylene dryer. You ran the calculations and have come to the conclusion that the residence time for the ethylene is too short. You have approached management with the great idea of a bigger dryer with more throughput, and it will only cost $1 million and 3 months of downtime. Management balks at the cost and downtime, so your next option is to alter the flows to and from the dryer to increase the residence time. The next step is to talk to Sam, the board operator for A Shift. Sam is well known for being a stickler for the rules and only doing what the code allows, and you know he won't like this idea.

What do you do? How do you get Sam to understand that this is going to make his job and his shift mates' job easier because they won't have to deal with over saturated ethylene being sent to other facilities downstream? What approach would be best to develop a collaborative relationship rather than a combative one?

Join the presenter, an experienced operator who is skilled at building effective cross-team relationships, as we discuss how to successfully navigate these difficult situations. We will explore how to appeal to non-engineering work practices, utilize your operations team's hands-on knowledge, and come to a mutual understanding of how everyone benefits when you work together.

638889 Representation in Education: Using Norbert Rilliuex's Work in Engineering Curriculum

Abstract:

Focuses on diversity in hiring and promotion have been somewhat successful at increasing representation in chemical engineering and in company management. However, often the chokepoint preventing people of color from becoming involved in chemical engineering and other STEM fields is in undergraduate education. Despite similar levels of initial enrollment in STEM fields, people of color have a much lower rate of completion than their white peers. There are myriad factors for this, including lack of peer support, lack of mentorship opportunities, bias from professors, and other challenges. One problem is the lack of representation of people of color in faculty and course materials.

In this work, we discuss the contributions of Norbert Rilliuex, a 19th century scientist, entrepreneur, and one of the earliest chemical engineers. Rilliuex was the son of a slave and her "owner", but traveled to France to receive a first-class scientific education. Using a sophisticated separations system, Rilliuex turned sugar from a luxury to a staple by improving the quality and quantity of sugar commercially available. Additionally, his work made sugar production far less dangerous for the enslaved people who worked in sugar production facilities. Despite his contributions to the field, Rilliuex's work has faded into anonymity. In addition to reintroducing his work, we discuss how Rilliuex's work can be used as part of thermodynamics, separations, or introductory engineering courses. In doing so, we hope to increase the representation of people of color in the classroom and foster a more inclusive environment in chemical engineering departments.

AUTHOR INDEX

AIChE
120 Wall Street, FL 23
New York, NY 10005-4020

CURRAN ASSOCIATES INC.
proceedings.com

ISBN 978-1-7138-6823-1
9 781713 868231